Ogi Orange
THE BIG RACE

© 2023 Lee Chau

ISBN 979-8-9875895-3-3

Published by Pybabi Publishing

This book is dedicated to my dear brother, two sisters and sister in-law. I thank you for always encouraging me to move forward.

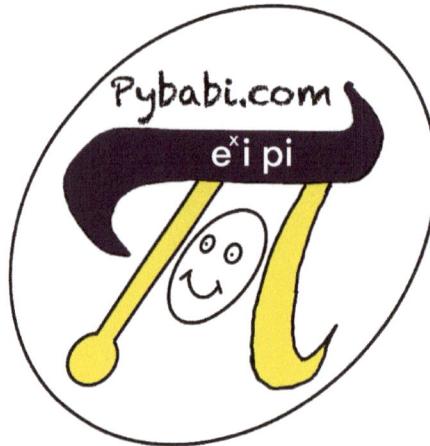

Message for instructor, teacher, or parent:

This book is best used by allowing one of the children who can read to lead the group after you have done it first as a model for them to imitate. Then they should read the story or section, ask the questions in the MATCH and ANSWER sections to their fellow students or if in a family to their brothers and sisters. As they lead the group you can monitor their progress. As the monitor you should answer any questions that arise naturally from the discussion. After one child has taken the lead, the role of teacher should be rotated to another child and then repeated until each child has successfully taught the material before advancing to another book. Since they are learning to teach the material they will naturally pay sufficient attention to learn the material and the lessons will stick with them. As the monitor, it may be best to intervene only as necessary allowing the children to solve and figure things out on their own until they need assistance.

If the child cannot read the parent can read and explain the lesson being taught. As soon as the child can understand even the simplest principle or idea have them to explain it back to you as if you are the student and they are the teacher.

When children feel they have an important role in education they will naturally learn what is necessary to play that role.

Of course this is not the only way to use the material, as you help the student you will soon discover which methods are best depending on the age and ability of the child. Please note that the illustrations in the book are specifically designed not to be perfect. They are only the beginning of a bigger idea or concept that will later be explained in greater detail. By keeping this in mind you will not have to push the student into trying to understand everything at the same time.

Thank you for choosing to use this material.

pybabi.com

OGI ORANGE

THE BIG RACE

Part I

Written and Illustrated by Lee Chau

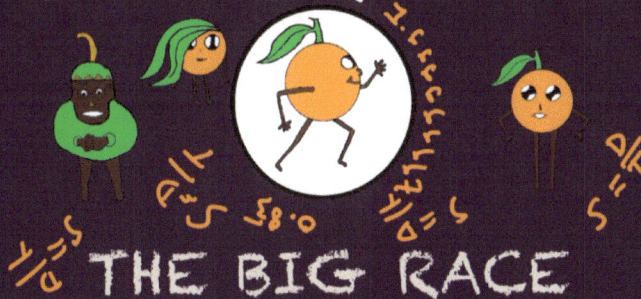

OGI ORANGE

THE BIG RACE

Part II

Written and Illustrated by Lee Chau

OGI ORANGE

THE BIG RACE

Part III

Written and Illustrated by Lee Chau

OGI ORANGE

THE BIG RACE

Part IV

Written and Illustrated by Lee Chau

This is my book

Date: _____

Name:

_____

I love you Space

Please sign, date and leave words of encouragement

I love you Space

Please sign, date and leave words of encouragement

I love you Space

Please sign, date and leave words of encouragement

Hey Secky & Disty, thanks for the lesson on speed, but what is 'THE BIG IDEA'?

Ok! Let's review!

Do you remember these lessons?

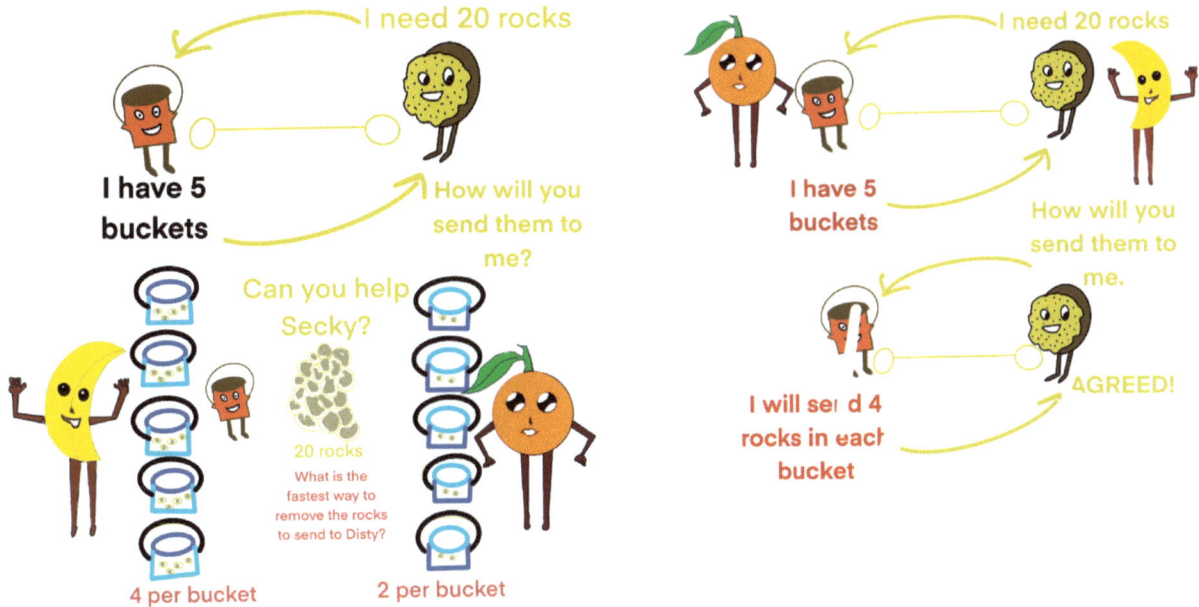

I need 20 rocks

I have 5 buckets

How will you send them to me?

I need 20 rocks

I have 5 buckets

How will you send them to me.

Can you help Secky?

20 rocks
What is the fastest way to remove the rocks to send to Disty?

4 per bucket

2 per bucket

I will send 4 rocks in each bucket

AGREED!

We use illustrations like these so that you can understand the principles. Once you understand the principles then you can break away from the illustrations or examples and keep the principle.

Let's use these lessons to show you how 'speed; distance; and time are connected together.

2

First let's come to an agreement. That is to say: make a ratio.

I need 10 rocks

If we break away from the example of rocks and buckets we can calculate our speed.

I have 5 buckets

$$\frac{\overset{\text{Rocks}}{\underset{\text{Distance miles}}{10}}}{\underset{\underset{\text{Buckets}}{\text{Time minutes}}}{5}} = \text{2 miles per minute}$$ Speed

3

We want to keep our speed the same but double our distance.

If we break from the example we can calculate our speed.

That means that we keep 2 rocks per bucket but I need 20 rocks instead of 10.

$$\frac{10 \text{ (Rocks / Distance miles)}}{5 \text{ (Time minutes / Buckets)}} = 2 \text{ miles per minute} \quad \text{Speed}$$

$$\frac{20 \text{ (Rocks / Distance miles)}}{? \text{ (Time minutes / Buckets)}} = 2 \text{ miles per minute} \quad \text{Speed}$$

What adjustment will I have to make with my buckets?

4

To keep the SAME speed after doubling the distance we need to double the time.

I need to use 10 buckets instead of 5.

$$\frac{\substack{\text{Rocks}\\\text{Distance miles}\\10}}{\substack{5\\\text{Time minutes}\\\text{Buckets}}} = \text{2 miles per minute} \quad \text{Speed}$$

$$\frac{\substack{\text{Rocks}\\\text{Distance miles}\\20}}{\substack{?\\\text{Time minutes}\\\text{Buckets}}} = \text{2 miles per minute} \quad \text{Speed}$$

$$\frac{\substack{\text{Rocks}\\\text{Distance miles}\\20}}{\substack{10\\\text{Time minutes}\\\text{Buckets}}} = \text{2 miles per minute} \quad \text{Speed}$$

Thanks!

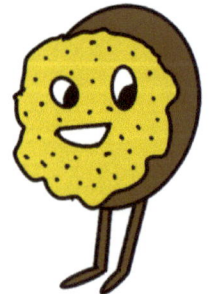

5

Remember! Our time does two things. It helps us with our 'rocks per bucket' and it is the total amount of time that it takes to get to our destination.

Your ETA (estimated time of arrival) will be in 10 minutes instead of 5 minutes because you doubled the distance but the speed is the same.

$$\frac{\overset{\text{Rocks}}{\underset{\text{Distance miles}}{10}}}{\underset{\substack{\text{Time minutes}\\\text{Buckets}}}{5}} = \underset{\text{minute}}{\text{2 miles per}} \quad \text{Speed}$$

$$\frac{\overset{\substack{\text{Rocks}\\\text{Distance miles}}}{20}}{\underset{\substack{\text{Time minutes}\\\text{Buckets}}}{?}} = \underset{\text{minute}}{\text{2 miles per}} \quad \text{Speed}$$

$$\frac{\overset{\substack{\text{Rocks}\\\text{Distance miles}}}{20}}{\underset{\substack{\text{Time minutes}\\\text{Buckets}}}{10}} = \underset{\text{minute}}{\text{2 miles per}} \quad \text{Speed}$$

Ok got it!

6

We can also see that if we keep the speed the SAME but cut the distance in HALF then the time will be cut in HALF also.

The less rocks that are requested the less buckets I need to use! I will only use half the buckets now. That is 5.

Rocks
Distance miles

$$\frac{20}{10} = 2 \text{ miles per minute} \quad \text{Speed}$$

Time minutes
Buckets

Rocks
Distance miles

$$\frac{10}{5} = 2 \text{ miles per minute} \quad \text{Speed}$$

Time minutes
Buckets

The shorter the trip the faster we get there.

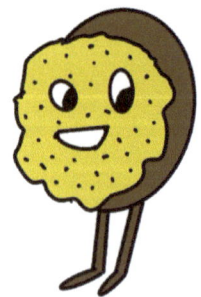

7

What happens if we keep the distance the SAME but double the speed? The time is cut in HALF!

The more rocks I put into each bucket the less buckets I need to send.

Rocks
Distance miles

$$\frac{10}{?} = 4 \text{ miles per minute}$$

Time minutes
Buckets

Speed

Rocks
Distance miles

$$\frac{10}{2.5} = 4 \text{ miles per minute}$$

Time minutes
Buckets

Speed

The faster we go the less time it takes to get there.

8

It also shows that if we leave the distance the SAME but cut the speed in HALF then the time will double!

The less rocks I put into each bucket the more buckets I need to send.

$$\frac{10}{2.5} = 4 \text{ miles per minute}$$

Rocks
Distance miles

Time minutes
Buckets

Speed

$$\frac{10}{5} = 2 \text{ miles per minute}$$

Rocks
Distance miles

Time minutes
Buckets

Speed

The slower we go the longer time it takes to get there.

9

If we cut the time in HALF and keep the speed the SAME. Then we HALF the distance.

The less buckets I can use with the same rocks per bucket the less total rocks I can deliver.

Rocks
Distance miles

$$\frac{10}{2.5} = 2 \text{ miles per minute}$$

Time minutes
Buckets

Speed

Rocks
Distance miles

$$\frac{5}{2.5} = 2 \text{ miles per minute}$$

Time minutes
Buckets

Speed

The less time at same speed we cannot go as far.

10

If we double the amount of time and keep the speed the SAME we will double the distance.

The more buckets I can use with the same rocks per bucket the greater the total rocks I can deliver.

Rocks
Distance miles
$$\frac{5}{2.5} = 2 \text{ miles per minute}$$
Time minutes
Buckets

Speed

Rocks
Distance miles
$$\frac{10}{5} = 2 \text{ miles per minute}$$
Time minutes
Buckets

Speed

The more time I travel at the same speed. The further I can go.

11

 # MATCH 1

What is an agreement?

What do the rocks represent?

What do the buckets represent?

It means that for every number of one thing The other will constribute a certain number of something else like 1 bucket for every 2 rocks.

Is time, distance and speed connected to one another?

Why do we use the example of rocks and buckets?

A ratio

Distance

To make things easier to understand and then break away from the example and keep what we have learned from it.

What are the normal units of time?

Time

Yes

What is another name for an agreement?

Minutes, Seconds, Hours

MATCH 2

What are the words we use to measure distance?

What is the symbol for the word 'per'?

A bar or line between the numbers

Miles, Kilometers

MATCH 3

If the distance doubles but our speed remains the same. What will happen to our time?

What is speed?

Time Doubles

Distance over Time.

MATCH 4

What are the two things that our time help us with?

How long is the total trip. Total distance.

How far we have to travel.

What does our distance tell us?

1. Calculate speed
2. Arrival time ETA

What is speed?

What happens if we keep the speed the same but cut the distance in half?

The time is cut in half.

Distance over Time.

What is speed?

What happens if we keep the distance the same but doubles the speed?

The time is cut in half.

Distance over Time.

MATCH 7

What is speed?

What happens if we keep the distance the same but cut the speed in half?

The time is doubled.

Distance over Time.

What is speed?

What happens if we cut the time in half but leave the speed the same?

Then we half the distance.

Distance over Time.

 # MATCH 9

What is speed?

What happens if we double the amount of time but keep the speed the same?

Then we double the distance.

Distance over Time.

ANSWER 1

What happens if we keep the speed the same but cut the distance in half?

What happens if we keep the distance the same but cut the speed in half?

What happens if we double the amount of time but keep the speed the same?

If the distance doubles but our speed remains the same. What will happen to our time?

What happens if we keep the distance the same but doubles the speed?

What happens if we cut the time in half but leave the speed the same?

ANSWER 2

Is time, distance and speed connected to one another?

Why do we use the example of rocks and buckets?

What do the buckets represent?

What are the normal units of time?

What does the rocks represent?

What is another name for an agreement?

What is an agreement?

Teachers Space

Please sign, date and leave words of encouragement

Friends Space

Mess up Space

Do whatever you want here.

OGI ORANGE
THE BIG RACE
For ages 2 years and Older
Part I
Written and Illustrated by Lee Chau

OGI ORANGE
THE BIG RACE
For ages 2 years and Older
Part II
Written and Illustrated by Lee Chau

OGI ORANGE
THE BIG RACE
For ages 2 years and Older
Part III
Written and Illustrated by Lee Chau

OGI ORANGE
THE BIG RACE
For ages 2 years and Older
Part IV
Written and Illustrated by Lee Chau

Let's go to Castle Calculus!

Can you share your story?

Tiktok @ogi.orange

Twitter @ogi_orange

Instagram @ogi.orange

Email: ogiorange@pybabi.com